Why the Wind Sings

Ngobile's Books

For every child asking questions and searching for answers...
Never stop asking.
Never stop investigating.
Never stop wondering.

Why the Wind Sings

Written By,
Dr. V

Sometimes I hear a whisper

It's music to my ears.

The branches are moving

and leaves dancing happily.

Can you guess what I hear?

I ask my mommy, "Who is singing?"
And she laughs and claps her hands.
"Oh that's not music you hear, my Love,
It's wind blowing across the land."

"How far does wind reach, Mommy?"

I ask while peeking at the sky.

"It reaches here, there, and everywhere

That's why the sand is whizzing by."

"What makes the wind sound loud sometimes?

So far away, but sounds so near.

I can feel it on my skin right now

And I certainly can hear."

"My Love, the sound comes from vibrations
as objects are touched by the moving air.
Things shake, move, wiggle,
and wave at the sun
Causing sound waves to flow
from here to there.
These waves crash into your ear drums,
while blowing through your curly hair."

"Some sound waves sing high and loud,
while others are gentle and low.
Loud wind is air moving hard and fast.
Quiet wind comes when it blows soft and slow."

I love to ask my mommy
about things I hear and things I see.
She tells me I'm inquisitive,
and explains everything to me!

"Well, Mommy, what is a sound wave?"

I ask excitedly

"That's a question for tomorrow," she says.

Then pulls me onto her knee.

Just to see if she's as smart as I think

I ask her something not so easy.

She laughs when she sees me scratch my head

But I know she's only teasing.

"I love that you ask hard questions,

and that you always want to know.

Keep trying, and reading, and learning

new things;

You never know how far you'll go.

The world is full of mysteries

just waiting to unfold.

You be the one to make discoveries

and tell the stories yet untold."